W9-ACL-715

What is my CAT thinking?

What is my CAT thinking?

GWEN BAILEY

CHARTWELL BOOKS

This edition published in 2016 by
CHARTWELL BOOKS
an imprint of Book Sales
a division of Quarto Publishing Group USA Inc.
142 West 36th Street, 4th Floor
New York, New York 10018, USA

ISBN-13: 978-0-7858-3430-4

Printed in China

First published in Great Britain in 2002 by Hamlyn,
a division of Octopus Publishing Group Ltd
Carmelite House, 50 Victoria Embankment
London EC4Y 0DZ
www.octopusbooks.co.uk

An Hachette UK Company
www.hachette.co.uk

Contents

Introduction

Cats are known and loved for their independence and free spirit. When they choose to interact with us, we often feel privileged and try to relate to them as best we can. We would all like to be able to speak their language so that we could ask them how they are or why they do what they do. However, unless there is a major scientific discovery or we learn to use telepathy, we will not be able to do so in our lifetime or theirs. So how, instead, can we learn what our cats are thinking?

Currently we do not have many scientific studies of domestic cats to back up the knowledge gained by experience. Perhaps we will one day, but until such time we have only our observations and experience from which to draw conclusions.

If we observe our pet cats carefully, we can build up a detailed picture of their body language and actions that helps us guess how they might be feeling. Through my work as an animal behaviourist and through living with cats at different times in my life, I have been lucky enough to watch many cats in different situations, and study photographs and video footage of cats showing extreme behaviour. I have seen how they appear to have similar emotions to ours, and to react in quite similar ways. If you add to this a knowledge of the species differences between us and a detailed study of their

behaviour, it becomes easy to guess what they might be thinking at any given time.

Acquiring sufficient knowledge to be able to do this takes time, so this book is an attempt to help you to learn what cats are saying so that you too can begin to guess at what they may be thinking.

I have tried not to be anthropomorphic and base my comments on what a human would think in the same situation. However, how I think will inevitably have an influence on how I empathize with others of a different species. Nor have I said what I think people want to hear or what might be cute, but instead what I really think cats would say if they had a voice.

I hope that you will be able to use this book to help you to work out what your cat and other cats you meet may be thinking. By doing so, you will be able to understand them better and give them a better life. Too often I see the results of cats being misunderstood by people: cats struggling to get along with other unrelated cats in a household because their owner likes cats and wants to keep several in a small territory; cats denied access to an unsafe outside world and bored or frustrated with nothing to do, or cats punished for natural behaviours that don't tie in with the owner's idea of how a cat should behave. Many of these problems could be prevented and cats given a better life if owners understood their pets better. Part of this involves an ability to read their body language and postures and to learn why they do certain actions. As intelligent humans, it is our responsibility to learn about the language of the animals we keep as pets – I hope this book will help you to make a start.

Social animals

Cats are not pack animals like dogs and it is remarkable that they have adapted so well to living with people. In feral colonies where groups of cats live wild around a food source without human owners, there is little interaction between unrelated cats and they only get close to each other when they have to. It is surprising, therefore, that they tolerate humans so well and even seek our affection. Their behaviour towards us is very kitten-like. We enjoy these behaviours and reward them with attention, affection and food, so cats perform them more often.

It's comfortable here

Even old cats are very mobile and will seek out the places where there are greatest comforts. This elderly cat will sleep most of his remaining days away, but likes to be where he is most comfortable. His owner provides a warm, soft lap, together with some affectionate stroking and grooming. In his own home, he knows he is safe and can relax and sleep deeply.

DID YOU KNOW?

- The genes a kitten inherits from his father will determine how friendly he will become. His mother's genes will also play a part but the effect is not so strong.
- Friendly mothers are more likely to produce sociable kittens as they show their offspring that there is nothing to be afraid of.

I like humans

This cat has been well handled and treated kindly since birth. He is trusting and enjoys the company of a complete stranger. Although he is not on his own territory, this cat copes well and seems to draw comfort from the people around him. He rubs his head on the person's arm to leave behind his scent. This begins the process of mixing the scents to produce a communal scent, which is important to group-living cats.

Great food!

There is good food at this hotel, which means that this cat is likely to stay. Cats are opportunistic and will go where the best food is. They don't have any real loyalty to the family as dogs do, but they usually stay because they bond to the home and territory where they live. If, however, food is scarce in one part of their territory, they will happily move next door where there may be a better supply.

Communal scent

Cats like to rub against us to mix their scent with ours to form a communal scent. They will do this with all familiar, friendly humans and animals in the family. The resulting communal scent is reassuring, smells like home and represents safety. Glands on the side of the mouth, around the face, on the head, along the tail and at the base of the tail all produce a special cocktail of scent chemicals which carry the cat's personal signature. Mixing their smell with ours is a vital function during social interaction and cats spend a lot of time rubbing against us.

On my head

Skin glands on this cat's head rub on the hand as she is stroked. This is rewarding for the person as the cat reciprocates the stroking as she bunts against the hand. It is rewarding for the cat since she gets to leave scent on the person, so they are both happy. Cats will often bunt against low branches or any other object that they pass underneath. Sometimes, a little jump with the front feet off the ground will be needed in order to reach.

Just here

This cat has rubbed her mouth along the outstretched fingers and finishes off by rubbing the other scent glands along her face against the hand. We don't know yet whether different messages are conveyed by the skin glands in different places on the body, but cats seem to show no preference and use whichever glands are closest to the item to be rubbed against.

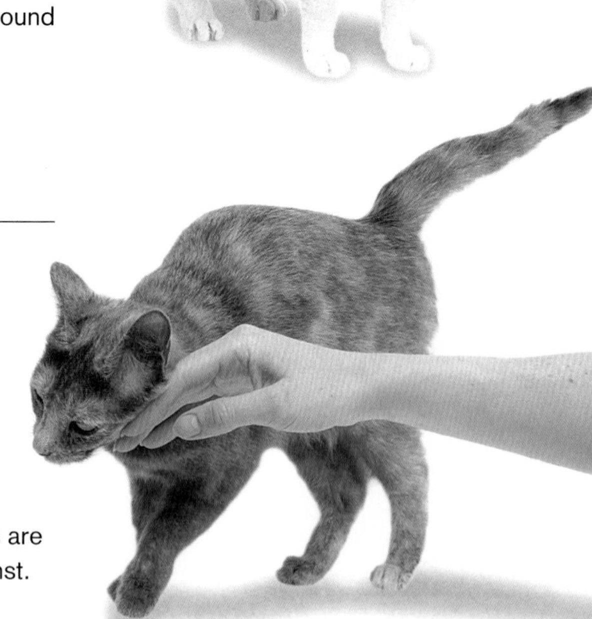

Hold still

This person's nose is level with the cat's head, so the cat uses the glands on the top of his head to leave his mark. As he does so, he closes one eye so that it is not damaged, and the person closes hers for the same reason. He seems to enjoy this and the person, who sees it as friendly attention from him, smiles her appreciation.

Keep going

As this cat is stroked along her back, she raises her back end up so that the skin glands at the base of her tail are pushed against the hand. Her tail is bent but stiff, as it usually is when cats are concentrating on spreading their scent on people or other animals.

DID YOU KNOW?

❑ A cat that has been to the vet may come back smelling very differently and may be treated like a stranger by other cats that once lived in harmony with him. Separating them and swapping their scents with a cloth can help.

Ah, that's nice

This cat is exerting a slight pressure as the hand passes her face and she seems to enjoy being stroked along her cheek, past her eye and over her ear.

Greetings

Anyone who owns a cat will know the tail-up greeting that cats give to someone they know. Usually accompanied by a special little trilling noise, cats will also show this greeting to other felines. If this signal is reciprocated by another cat, it is often followed by rubbing head against head, followed by a body rub against the flanks of the other cat. Since cats are not very tall, their greeting to us is usually followed by weaving around our legs as they try to rub against us.

Hey, you're home

These two cats are greeting their owner as she returns home. Their tails are held vertically in the classic cat signal of greeting. The base of the tail is held upright and rigid while the tip is allowed to bend over. Their owner reciprocates in a human way by smiling and reaching out to touch them. The cats will be familiar with this but would understand better if she had a tail she could raise up in the air.

Hello stranger

Some cats are so friendly that they will approach total strangers in the hope of getting some social contact. This three-legged cat approaches a stranger with his tail raised in greeting although with slight apprehension about how he will be received.

DID YOU KNOW?

- The amount of socialization a kitten has with people while he is between two and seven weeks old will determine how well he will interact with people later in life. Good experiences in early life will produce a friendly, outgoing cat.

Nice to meet you

Having discovered that the stranger is friendly, the tail continues to be held up while the cat rubs his face and head against her hand. Cats that have been through traumatic experiences, such as a car accident, go through a second socialization phase while they are still hospitalized which can mean that they become much more friendly with people than they once were. Sadly, some can go the other way and become less friendly than they were previously.

Stroking

Cats show very kitten-like reactions to being stroked by us. They begin to purr, and knead and tread with their front paws. Both of these actions have their origins in kittenhood, when the kittens would have kneaded with their paws to stimulate the mother to let down her milk, and purred contentedly with her once they were suckling. When they are enjoying human company, they are likely to show the characteristic kneading and purring behaviour that symbolizes a contented cat. Our stroking is similar to the actions of a mother cat when she is licking her kittens and causes them to return to the behaviour they displayed early in their lives.

I'll just have a snooze

This cat is used to being cuddled as this has happened since he was a kitten. He knows and trusts his owner and is content to accept being restrained in this way. Other cats may find this situation a bit claustrophobic and may wriggle loose or try to fight their way free in panic. This cat, however, shuts his eyes and relaxes in the knowledge that he is safe and his owner will take responsibility for his welfare while he is with her, as a mother cat may do for her kittens.

I'm so relaxed

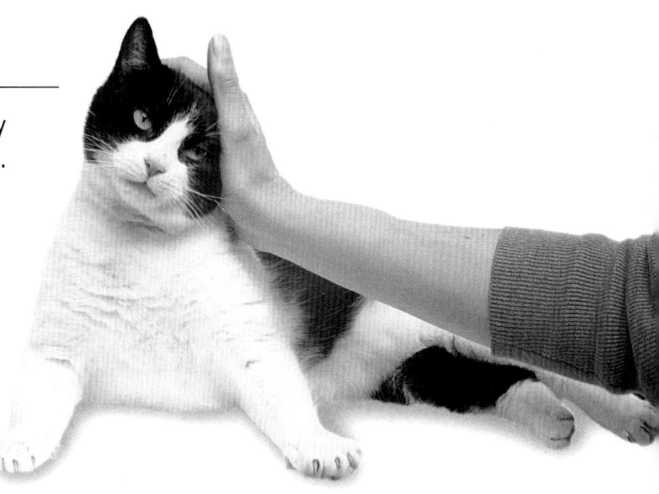

Relaxed cats stretch out contentedly and this cat is enjoying the attention. He is very friendly with people and enjoys their company. As he is stroked, he purrs loudly, making a noise continuously as he breathes in and out. He goes through the rhythmic treading motions and here you can see the left paw has its claws extended and is pushing against the ground.

I don't like this as much

Here the right paw has its claws extended and the left paw's claws are sheathed. If he was sitting on your lap this cat might be pushing his paws down, which could result in his claws digging into your skin. However, the hand is now covering his ears and the top of his head and he looks a little disgruntled.

DID YOU KNOW?

- Different cats dig in their claws while being stroked to different degrees, but this is a natural behaviour and it is not possible to train them not to do it.
- Cats will not understand if you punish them for digging their claws in. Placing a thick blanket or a towel underneath them when they sit on your lap can help to prevent any discomfort, as well as keeping any stray hairs off your clothes.

Different social requirements

Different cats require different degrees of social contact and some will be more independent and aloof than their owners wish. Cats that have a low requirement for social contact can make disappointing pets, since although they may learn to tolerate their owners' attention, they never really seem to enjoy it. Other cats will actively seek out human company and show signs of distress if they cannot get enough. Cat breeds such as Siamese and Burmese are very popular because selective breeding has resulted in a high need for contact with their owners.

Leave me to sleep

Cats often tolerate our attempts to be friendly, but don't always appreciate it or even seem to enjoy it much. This cat is well socialized and very trusting of humans, but would probably like to be left alone to sleep. He lifts his head slightly as his owner's hand runs over the top of his head in a token effort to rub himself against her, but he soon gives up and settles back down to sleep.

DID YOU KNOW?

- ❑ Cats start to purr when they are one week old.
- ❑ They can purr continuously as they inhale and exhale.
- ❑ Purring is usually a sign of contentment, but cats also purr when they are in pain.

Are you friendly?

Some cats enjoy the company of people very much and this is one of them. She is waiting in a rescue centre for a new home and isn't sure about her visitors' intentions. She is weighing them up, half-raising her tail in greeting and has raised her paw ready to come forward if she sees signs that it is safe to approach. Once she has approached the people, she will rub herself against them and wind herself around their legs in an effort to make friends quickly.

I don't need you

Some cats are friendly for brief periods but spend most of the day on their own doing other things. Owners who have a high need for social contact with their cats can tire of them quickly unless they realize that they are like this because of their inherited personality, and try to have different expectations of them. Some cats stay away from humans because of a lack of trust due to poor socialization during kittenhood. These cats are timid and can usually be encouraged slowly to be more friendly with gentle treatment.

Cats and other animals

Being descended from animals that lived solitary lives, cats have few strategies for dealing with unrelated animals or those they do not know. How well they accept strangers and newcomers to the home depends largely on how well socialized they were during kittenhood and how many pleasant encounters they had with other animals of that species at this time. It also depends on how slowly they are introduced to new arrivals, how easily they can get to safety and whether they have sufficient time to learn to trust them before they get into close encounters.

I'll get you

These cats have grown up together and are happy to play. Their social play mimics real fights but both know it is not serious. Their claws are not out and their bites are inhibited. However, they posture and paw at each other as in a real fight and sometimes leap on one another, rolling and raking with their back feet to make the other back off. These play-fights help maintain their relationship and enable each to get to know the abilities and strength of the other.

Go away!

The tabby-and-white cat is not playing. She is trying to make the ginger cat back away and is doing so quite successfully. The action is rapid and sufficiently threatening to make the ginger cat take avoiding action. However, these cats belong to the same household and know each other well, so there is no need for real aggression. The tabby-and-white cat keeps her claws sheathed.

It's warmer here

Having grown up with this gentle dog and learned to trust him, this cat is quite comfortable to lie down in his bed. The dog is happy to share his bed and poses no threat to the cat. The cat knows that by snuggling up to the dog, she can get some extra warmth that is great on a cold day. Trust like this would be hard to gain with a new dog that the cat hadn't grown up with, but it may come in time if the dog proves to be safe enough.

Do I know you?

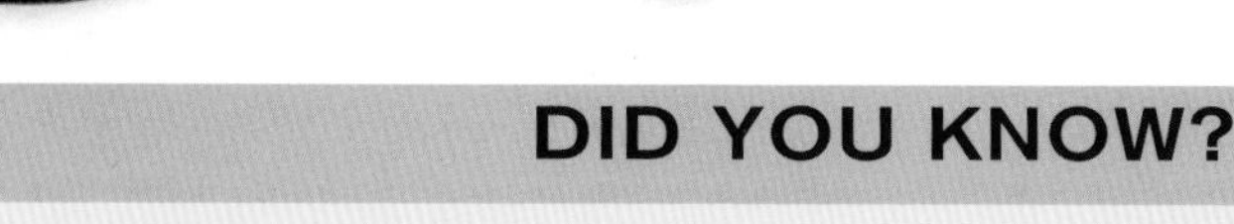

Nose-to-nose greetings like this are unusual as they put both cats in a vulnerable position. However, this is mother and daughter and, although they have not seen each other for a while, they know each other well and feel safe. Sniffing in this way helps each to confirm their visual recognition and to gather more information about how the other cat is, where they have been and what they have been doing.

DID YOU KNOW?

- ❑ Adult cats are more likely to adjust to a new kitten than to a new adult cat that is introduced into the household.
- ❑ Newcomers such as other animals or even babies and new people can result in the cat feeling insecure.
- ❑ Cats that feel their home security has been breached may spray urine or scratch in strategic places to make themselves feel more secure.

Body language and signals

Apart from the tail-up greeting, other visual social signals are subtle and hard to read apart from those needed for defence or aggression. This is a legacy from cats' solitary ancestors who had little need to communicate with each other as they lived their lives alone unless they were fighting or reproducing. However, cats will react to certain situations with different body postures, and by studying their body language we can tell how relaxed or tense they are or guess at what they may be thinking.

Get me out of here

This cat is anxious and worried. His tail is tucked under to keep it out of the way should anything attack him, and his weight is mostly centred over his hind legs so that he is ready to run or strike with his front claws if necessary. His ears and whiskers are rotated to keep them out of the way in a fight, and his eyes look upwards to find out if there is a safer place higher up that he can reach. He is meowing loudly, hoping that his owner will rescue him.

I'm safe

In contrast to the black cat, this one is totally relaxed. He lies down with his hind legs outstretched and vulnerable. His front paws are curled up in a comfortable position. Since front paws with their claws are very useful in defence, curled paws like these show that this cat has no fear that they will be needed. His face is relaxed, but something has caught his interest and he pricks up his ears and looks in that direction to pick up any useful information.

Anybody want to stroke me?

This cat is totally relaxed and feels secure enough to stretch out in a very vulnerable position even when the dog is going past. Lying on their heads like this and waving their front paws is usually irresistible to owners, who can't help stroking them and giving them attention. This cat lives with lots of other cats and, although their owners have lots of love to give, she needs to put some effort into getting attention from them. Lying on the table in full view and adopting a cute pose works almost every time.

Territory

The ancestors of our domestic cats did not hunt in packs but were solitary hunters who patrolled a territory that supplied them with all the food they needed to survive. Although our pet cats are well fed by us and no longer need to hunt, they have not lost the desire to stake out an area they can call their own. To them, a territory represents safety and a food supply should their humans fail to provide. We may find this hard to appreciate but for cats, territory is sometimes more important to them than the family with whom they live.

I live here

A large hunting range is necessary if the cat is to catch the prey it requires to survive. Our domestic cats can afford to share their territories since we feed them all they need. Despite this, a domestic cat that has access to the outside will patrol a large hunting range and check it regularly to find out who has been there and what prey it may contain. This cat surveys the scene, checking for intruders. He will cover a lot of ground while out hunting and gets to know his territory intimately.

Keep your distance

In our overcrowded urban neighbourhoods, cats are required to tolerate the presence of others on a more frequent basis than they would if they were living wild. They avoid strange cats wherever possible, making use of resources on a time-share basis. If these two were dogs, you would expect them to go over, get to know each other, and eventually form a pack. Cats keep their distance and avoid each other.

My territory

Keeping a regular check on your territory is important and is easily done if you are high up. Cats do not cope well with changes in their territory, and find it very difficult to adjust to a new one when they move house. There are many stories of cats returning to the house where they used to live rather than staying with their family in their new home, particularly if they have not moved far away.

DID YOU KNOW?

- ❑ Cats are extremely good at making accurate mental maps of their territory.
- ❑ It takes time to know the territory well and this is why cats are so unsettled for the first six months in a new house.

What's going on?

A tree like this one makes a natural vantage point for cats, who can climb into them easily. This cat looks intently at something in his garden. It could be another cat that may need to be avoided or seen off, or a bird that could be hunted. It is important for cats to know exactly what is happening in their territory so that they can defend it, hunt in it or rest in it successfully.

Scent messages

A cat's sense of smell is very important to them. Compared to us, their ability to detect odours is incredible and we cannot begin to imagine the depth and variety of information they must be able to gather from just one sniffing session.

If, like a cat, your ability to communicate with body language is limited, getting too close to other cats and unknown animals is dangerous. Therefore, it makes sense to keep your distance and communicate at long range. Scent is an ideal mechanism for this purpose, as scent messages can be left that will linger in the environment for some time and inform others of your presence.

Who's been here?

Another cat or even a different animal may have passed by, leaving its scent on the edge of this tree stump. This cat sniffs carefully and for a long time, to take in all the subtle aromas. Getting close to the source of the scent is important and ensures she receives the information before the scent's more volatile components evaporate. This means she gets the complete message rather than a diluted version with bits missing.

DID YOU KNOW?

- Scientists have been able to isolate the chemical common to all cats that is rubbed on surfaces from the cheek glands. It is now reproduced synthetically and sold commercially to help make cats feel more at home in a new or disrupted environment.

Have some scent

Leaving your scent message on your owners is as important as leaving it on the non-living parts of your environment. This cat rubs his head and then his flanks and tail along his owner's leg to swap their scents. By doing this he shows that he recognizes her as part of his social group and he gains more security by having a communal scent that encompasses a number of individuals.

I'll just mark this

This cat wipes his cheek on a prominent stalk that is sticking up at just the right height in his garden. He is leaving scent from the skin glands on the side of his face on the stalk for others to read. Before doing this, he carefully sniffs at the stalk, checking to see if his previous mark is still there and whether it has been overlaid by the scent of another cat that has passed through his territory.

More information

Cats deposit scent marks to leave a reminder of whose territory it is, to make themselves feel more secure, to time-share a well-used area, or to advertise their presence.

As well as rubbing their scent glands on objects in their territory, they can also scratch suitable surfaces so that the scent glands between their toes are wiped on the surface, or can spray urine onto prominent landmarks. In exceptional cases, and if they really want to make a statement, very confident cats will leave faeces in prominent places (a process known as middening).

This is my area

This cat is spraying a prominent object in his garden. He has backed up to the object, raised his tail and then sprays urine while quivering his tail and paddling with his back feet. He has a look of concentration and his ears are slightly back. Only a little urine is sprayed, compared with the large quantity that is expelled when cats squat down to go to the toilet. Spraying leaves a scent message at the height of other cats' noses, which allows it to be read easily. It may also contain extra scents from the glands underneath the tail.

DID YOU KNOW?

- Insecure cats that feel their territory is threatened may spray in strategic places around the home. This fills their territory with their scent and helps them to feel more secure.
- Things that cause a cat to spray in the home can range from smells brought in on shoes and bags, to being bullied by another cat whenever they step outside the cat flap.
- Owners are often oblivious to these causes and punish the cat for being 'dirty' in the house, which further adds to his insecurity.

See and smell

Cats have scent glands between their toes that are activated during rhythmic stropping as they scratch a suitable surface. Although cats may do this to keep their claws operational, they also do it to leave messages for other cats. In multi-cat households, scratching may also serve as a visual signal of confidence or as a statement of identity. Well-used places often show signs of significant damage and such sites are often investigated by other cats in the area. This cat looks as though he is scratching the tree but his body is at the wrong angle and, if you look closely, you will see the piece of grass he is trying to catch!

Whose is this?

If an object is not thought to be dangerous and warrants further investigation, the nose will be called into action to find out more. This cat has come across a toy that has been used by other cats and she gets close with her nose to find out who has been there before her. While she concentrates on this, she keeps her other senses tuned outwards on her environment so that she will have an early warning of any impending danger.

Leaving and receiving

As well as informing others of their presence by leaving scent messages, cats can find out a lot about the other cats in their territory by reading the odour messages they leave. Although we do not know exactly what they can find out, it is likely that they can identify which individual in their territory left the scent mark. In the cases of spraying and middening, they are also likely to be able to tell how old the cat is, what sex it is, what state of health it is in and what it has eaten recently.

It's me

If you were a cat walking into this room, the sofa is one of the first things you would encounter. For this reason, it has become a favourite place for leaving scent marks. It bears the scars of previous scratching, but this cat, after carefully sniffing, chooses to rub his chin and face and, eventually, his whole body and tail along it. By doing so, he leaves his mark on a prominent place in his core territory and mixes his scent with those of the other cats that live there.

It's him

Another cat in the same household sniffs at the scent that has been deposited. After carefully examining it for some time, he adds his own. It smells familiar, he recognizes the cat that left it, and there is no animosity between them. If there was, a stronger message might be required and this cat may choose to scratch or spray to cover the scent of the intruder.

It's my scent

By sniffing at the scent on this chair, this cat can tell that he was the one who left it some time ago. From this, he will know that it is likely that no other cat has passed this way recently. His own scent is familiar and makes him feel safe in his territory.

I'll add some more

Scent is made up of chemicals that decay gradually in air. The scent of the residual components encourages cats to overlay the mark with more of their own scent. In this way, prominent marking places let other cats know of the continued presence of the occupant.

DID YOU KNOW?

- The area in the nose for detecting scent is ten times larger in cats than in humans. They also have a correspondingly larger part of the brain to help them decipher these scent messages.
- This increased ability to detect scents and the greater degree of complexity of information received accounts for the large amount of time cats spend locating and reading scent messages left behind by others.

Other cats in the territory

For cats that don't form friendships with others easily, coping with cats that are not part of your household but share your territory can be difficult. One way of dealing with overcrowding is to time-share the facilities by leaving scent messages advertising the fact that part of the territory is temporarily occupied.

Since chemical messages decay gradually, the reader of these messages will be able to tell how long ago that cat passed by. From this, and from a knowledge of the other cats in the area, a decision can be made as to whether it is safe to proceed or whether it may be better to wait for a while or go in a different direction.

This is my patch

Disputes between cats over territory are common in overcrowded urban areas where gardens are small and cats numerous. In such conditions, it is inevitable that face-to-face encounters occur occasionally. Cats usually deal with these by slow-motion posturing, staring and growling until one of them backs down and moves away. The winner is likely to be the one that is on his own territory, but after several encounters both will know which cat is stronger and is more likely to hold his ground.

DID YOU KNOW?

❑ By sniffing a tom cat's urine, female cats can tell whether or not he is a suitable mate. They are able to detect by-products of his food, which tell them how much fresh meat he has eaten and, hence, how good a hunter he is. This helps determine how suitable he would be as a father for her kittens.

Don't come closer

These two cats have reached a standoff as both are unwilling to back down. The cat that is standing is probably in a stronger position and may be the owner of the territory, which will give him added strength. Cats can sometimes hold positions like this for hours before one finally decides to withdraw. Once this happens, they do not have a large number of signals to let the other cat know that they have surrendered, so retreat must be very slow to prevent a chase and attack by the victor.

So he went by

Branches and other prominent points in the territory will be marked at head height for other cats to sniff. This cat has detected cat odour on this branch and stops to sniff carefully. Getting to know the smell of other cats in the area is important so that a picture of who goes where at what time can be built up.

Indoor or outdoor cat?

Car accidents, dog bites, cat fights, attacks from cat haters and theft are just some of the dangers that lurk for our cats outside the safety of the home environment. Despite this, many people choose to allow their cats to go outside to explore and fulfil their natural instincts. Having the freedom to choose and the ability to hunt, explore and exercise is undoubtedly beneficial for them, and owners have to weigh this up against the potential hazards that this freedom inevitably brings.

I can't run away

A medical condition prevents this cat from having complete freedom, but his owner tries to make up for his confinement by walking him on a harness. Cats can learn to tolerate this, especially if it is started at an early age. However, unlike dogs, cats do not cope well with lead-walking in busy areas because their usual reaction to threats is to run, hide or get up high. A harness and lead prevents this and can cause panic and considerable distress if the threat is great.

Is he out here?

Some cats may decide not to go outside if they are bullied or frightened by other cats or other experiences in their environment. This cat surveys the area outside the cat flap to see if he is safe before venturing further afield. It takes courage to have the confidence to go out again after an attack. Having his owner accompany him for the first few visits can help considerably.

I'm bored

Cats kept permanently inside need lots of things to do to keep them occupied, otherwise they can become lethargic and stressed. Games with toys, cat activity centres, and logs and walkways to climb can all help to provide indoor cats with some much needed exercise and play. Constantly introducing new variety into their lives in the form of new objects to explore and new games to play helps to keep them interested and alert. Keeping a cat indoors is not an easy option for owners, who need to put in a lot of extra effort in order to keep them entertained and content.

DID YOU KNOW?

❏ Rogue cats that bully others can cause local cats to stop venturing out. This can lead to toileting problems, depression and other behaviour problems. Agreeing a time-share system between neighbours for letting cats out can help.

I wish I could be out there

Cats that have been allowed outside in the past can find it more difficult to tolerate being confined. They may watch through the window and chatter their teeth at the sight of birds as if frustrated or perhaps going through some of the motions of catching them. Cats that have lived all their lives indoors do not know what they are missing, but instinctively find the outside world fascinating and stimulating.

Exploration

It is important for cats to know their territories well so that they can determine where the best hunting sites are located. Cats will explore rigorously, checking every last detail of their environment and finding out about anything new. Checking frequently is important as cats like to regularly make sure that all is well with their world. This accounts for their insistence that we open doors for them, only for them to take a look, see that all is well and to be content. Owners often find this behaviour difficult to understand and become frustrated that their cat does not go out through the door that has been opened for him, or wants to come back in again almost as soon as he has gone out.

I can find out

Whiskers are a great asset when exploring an environment. These long, thickened hairs around the mouth have many receptors at their roots that enable your cat to detect how large an opening is, to 'feel' his way in very dark conditions and to detect air currents strong enough to cause the whiskers to bend. As well as whiskers, cats have long hairs growing from just above the eyes that are equally sensitive to movement.

DID YOU KNOW?

- Cats explore vigorously to get to know a new territory well, and build up a detailed map inside their head. Using this map, they are able to find new routes home even if they have never been that way before.

Who are you?

Finding out who else is in your territory is essential. This cat watches the photographer while keeping a low profile and sniffing at scents left by other cats on the bushes. Being curious about what is happening helps your cat to build up a detailed picture of his environment. Knowing who his competitors and rivals are is important, and finding out if food stocks are running low may be vital in the fight for survival in the wild. Although there is no need to do this in their cosseted domestic world, cats have not lost the inborn desire to do so.

I can just get through

Cats can squeeze through amazingly small gaps. This cat appears to have been squashed in the door, but is just squeezing himself through the small opening at the side. Whiskers on the face are called into play to determine whether or not the hole is big enough to get through, and if the head and shoulders fit, the rest of the body will follow.

Curiosity and the cat

The saying 'curiosity killed the cat' appears at first reading to be true. Their explorations can lead them into dangerous situations and we often hear reports of cats being accidentally locked inside sheds and washing machines. However, for wild cats the opposite is probably true, since their curiosity provides them with an increased knowledge of hunting sites that often keeps them alive in times of food shortages. Although we now provide all the food a pet cat needs to survive, they have not lost their desire to explore and find out about the world in which they live.

All's well

This cat has investigated the appearance of a ladder in his territory and now uses it as a vantage point from which to survey the rest of his garden. Cats are very curious about new objects and they will all be investigated thoroughly to make sure they are not harmful, or to find out if they could be a good place to rest or useful in the continuous quest for food and hunting opportunities.

DID YOU KNOW?

- Cats have been known to travel many miles to get back to their familiar territory if their owners move house and let them out too soon.
- Conversely, cats have travelled vast distances to their owner's new home after being left behind. No-one knows how they knew which direction to take or how they knew where their owners lived.

Let's just see what's in here

This cat activity centre has a small opening that is just the right size to entice this cat to explore. If it were bigger, he would be able to see inside and might not bother to go in.

In I go

The cat can't resist a look and tucks himself into a space that is only just big enough. In the wild, places like this may provide good cover for prey and would be worth checking out.

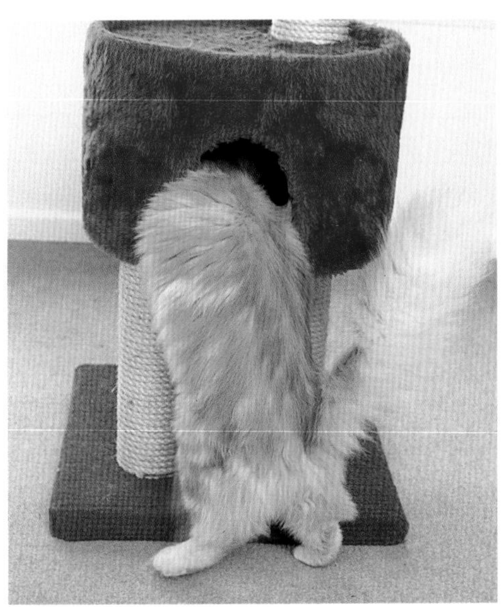

I've been here before

This cat has explored this place many times and found nothing, but he is always hopeful that there might be something of interest inside one day.

Fitness and hunting

Finding and catching prey is a basic instinct in cats. It is not necessary for them to be hungry in order to do so, although a hungry cat is likely to hunt more. Consequently, our domestic cats cannot help hunting even though we provide them with all the food they need. They are driven to do it by their instincts and for many pet owners catching and killing wildlife is one of the less appealing aspects of their personalities. Some cats are more avid hunters than others, but most cats do it to some degree. Although we can give them alternatives to hunting live prey in the form of games with toys, it is not possible to stop them doing it altogether unless they are kept permanently indoors.

You can't see me

This cat is stalking a toy and getting ready to pounce. He crouches low and keeps still with all his senses trained on the 'prey'. If he were outside hunting real prey, this strategy would be unsuccessful since he would be seen by his intended victim long before he could reach it. A surprise attack from undergrowth would make his ambush much more successful.

DID YOU KNOW?

- Cats prefer to hunt during dawn and dusk when small mammals are active.
- Cat's eyes are specially adapted so that they see better at lower light levels, to enable them to hunt successfully when the light is dim.
- Cats prefer to hunt small mammals rather than birds if they can find them. To catch them, cats wait patiently beside the runs and passages that they use.
- If out hunting during the day, only birds may be available. To catch birds, cats conceal themselves in undergrowth and wait for the birds to land.

I'm ready

Cats like to wait concealed in undergrowth ready to pounce on unsuspecting prey or birds that land on the ground. A basket makes suitable alternative cover for this cat, who bunches himself up ready to spring. He is completely motionless, with his eyes and ears focused on his toy 'prey'. He chooses his moment for the pounce carefully, waiting until the toy is near enough to grab easily.

Coming to get you

Cats are not great runners and ambush their victim from a convenient hiding place rather than running it to a standstill. The toy is still some way away but the cat has been unable to resist a pouncing movement, so the front legs have reached the ground before the back ones.

Too quick for me!

With his head held still so that all his senses can be kept fixed on the toy, he runs forward to swat it with his paws. If the toy had been prey, it would have been long gone by the time he caught up, and he would need to learn to wait until it was closer next time.

Pouncing and pawing

The first strike against a prey animal comes with the paws. Biting is saved for later, when the prey is already dazed and less able to fight back and cause injury. Pouncing raises the body up so that its weight can be used through the paws to stun the prey. Once the prey has been stunned, the paws and claws are used to prevent the prey escaping and to get it ready for the killing bite from the teeth. Cats usually kill fewer birds than rodents, but when they do, they use the same strategy. Once the bird is stunned and released, any effort to escape into the air is prevented by the cat rearing up so that the paws can reach it as it tries to flutter away.

Here I come

This is a spectacular pounce, showing this athletic cat in full flight. Hearing is as important as seeing in dense undergrowth and cats locate small rodents by their high-pitched squeaks and rustles as they move through the plants. Having located the victim using pricked ears to pinpoint the source of the sound, the cat leaps up and across to land with perfect precision on top of the prey with his front paws.

Missed

Kittens improve their instinctive skill at catching prey by oractising on anything that moves. This kitten has just missed the ball as it whizzed past. As his skill develops, he will learn not to waste energy on strikes that are likely to end in failure and will reserve his efforts for those with a much greater chance of success.

DID YOU KNOW?

- The sharp hooks on the ends of their claws allow cats to grip their prey.
- These hooks can snag in clothes and cause panic if cats cannot free their paws.

Let me get you

Rearing up to full height allows the cat to raise his paw high enough to catch a fluttering bird. His senses are focused on the prey and he lifts whichever paw is nearest to try to catch it. He still may not be high enough if the bird is fast, but he relies on his previous weakening of it and on his speed of reaction to make contact.

Now you're mine

This kitten practises his skills on this toy that flutters and bounces on the end of a line held by his owner. The best toys mimic the real thing and kittens are instinctively drawn to toys that have soft fluttering parts attached to a body about the size of a real bird. The kitten has raised his paw high and is ready to bring it down with claws extended in a fast strike, intended to catch the toy and flatten it beneath him to prevent its escape.

The catch

Once the prey is captured, it may need to be weakened by repeated release and capture if it is large or aggressive. Although the paws and claws do the capturing, the kill is made by a lethal bite during which a canine tooth is inserted in between the vertebrae, which dislocates the unfortunate creature's neck. Weakened prey are less likely to retaliate and injure the cat's face. The whiskers and sensitive hairs on the mouth and lips play an important role in this process and sensors at the root of the cat's canine teeth also help to fine-tune their placement during the kill. Cats are beautifully designed hunting machines and, as distasteful as it may seem to pet owners, killing is something cats are programmed for and it is unfair of us to expect them not to do it.

DID YOU KNOW?

- ❏ Cats often appear to be playing with their prey as they injure them a little, let them go and recapture them.
- ❏ 'Playing' with prey is probably due to a lack of experience, causing them to be poor killers.
- ❏ 'Playing' can also be a way of weakening prey that may otherwise bite its attacker in self-defence, before the cat puts its face close enough to dispatch the victim.

What are you?

Caution is sensible if you are not sure whether your intended victim will strike back. This cat is a very good hunter, but when faced with a strange new toy she is not too sure whether it is safe. Its movement has attracted her towards it, but she approaches carefully, ready to recoil quickly if it gets tricky. She uses her paw to pat at it to see if it will move so that she can make an assessment of what it will do.

Stay here where I can get you

Having 'caught' this toy mouse, the cat holds it between his mouth and front paws and is using his back feet to rake at it. This is play, as the cat would not let live prey get so close to his vulnerable belly region. However, it is useful practice for fights with other cats or defence. Although the back claws are often less sharp than the front ones, the hind leg muscles are strong and, if used in earnest in this way, can cause significant damage to the underbelly of an opponent.

Got you

This cat is relaxed and playing, but he is an efficient hunter when outside and knows what to do to dispatch prey. If he were positioning the 'prey' to deliver the killing bite with his canine teeth, he would be on top of it, pinning it to the ground.

Agility

Cats have the amazing ability to leap gracefully up onto very high places, to balance precariously on thin ledges and to right themselves quickly when they fall. In the wild, these abilities are essential if they are going to find enough to eat and to stay safe. To enable them to do this, they are well equipped with special apparatus and connections within the brain and body that tell the cat exactly what his body is doing at any given moment. Reflexes that do not require conscious thought help increase the cat's speed of reaction and enable him to react more quickly than he can think.

Going up

Cats have strong hind leg muscles and, if fit and well, find it easy to jump up onto very high ledges from a standing start. This enables them to get out of the way of danger and allows them to gain access to territory that would otherwise be denied by high fences. Sensors on the face and paws let them know when they reach their destination, and their finely tuned sense of balance enables them to land precisely where they want to.

DID YOU KNOW?

- ❏ A fit, healthy cat can jump up to five times his own height.
- ❏ Cats have special vertebrae in their spines that allow them to twist and flex much more than they would be able to with a more rigid backbone.

Where's that bird?

Cats are always interested in small moving creatures, even when these are completely out of reach. This cat is excited by the prospect of catching birds and follows them across his territory, watching their habits and looking for a chance to catch them. Birds will often mob cats, especially during their breeding season. Despite being well armed, cats will not risk being injured by dive-bombing birds and rapidly retreat to a safer place.

Going down

This cat does not have much room to manoeuvre and so cannot land on all four feet. Cats have very sensitive feet and have a few thickened hairs at the back of the wrists that will help to let this cat know when he is reaching the table top.

Displaced hunting

Some cats have a stronger instinct to hunt than others. If they are kept indoors with no opportunity to do so, they may turn their attention to other things that move in the house. Adults, children and other pets are often the target of their 'attacks'. Sometimes mock chases and bites will be just social play and, in these cases, the bites will be inhibited. However, in some cases, real predatory behaviour is seen and, since this is designed to catch and kill prey, considerable damage can be done to the victim. Giving the cat the opportunity to go hunting outside or to play with and 'kill' toys is part of the solution to the problem. However, the fact that cats will do this to members of the family with whom they are usually very friendly shows how 'pre-programmed' hunting behaviour is.

Surprise!

This cat lies in wait for his owner as she comes into the hall, and pounces onto her moving legs and feet as if she were a prey animal. Fortunately, he has plenty of opportunities to play and can also go outside, so these 'attacks' have more elements of social play than real predation and, as such, his bites are inhibited and his claws remain sheathed.

DID YOU KNOW?

- ❑ Some cats have stronger predatory instincts and a stronger desire to hunt than others.
- ❑ Keeping voracious hunters entertained at home with mock chases and capture of toys can help to prevent them going out to hunt. However, once outside and surrounded by opportunity, hunting will be back on their agenda.

Going to get it

Kittens play to perfect their skills as catching moving objects may be essential to their survival later on. This dog's tail provides a tantalizing moving object that periodically wriggles on the ground like a real prey animal. If this kitten had a very strong instinct to hunt and was denied the opportunity to do so later in life, he might well turn his attentions to an alternative moving object like this.

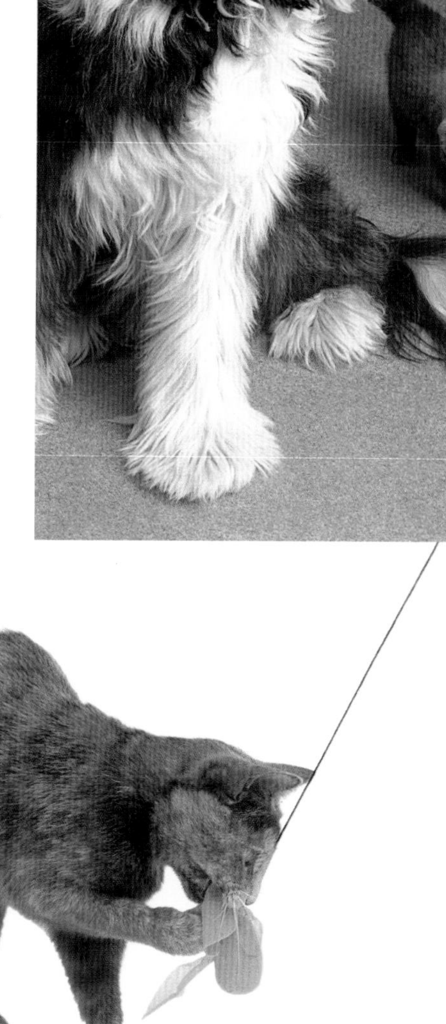

Got it!

Having a strong hunting instinct, this cat is easily excited into a game with a toy. Allowing her to 'hunt' it and catch it occasionally provides her with an outlet for her hunting energies and, after several small sessions a day, she will be less inclined to go out and catch wildlife. The more the owner can cause the toy's movements to resemble the behaviour of real prey, the better the cat will like it and the more contented she will feel after a play session.

Bringing home the catch

Cats in the wild will carry their prey to cover, and eat it in relative seclusion. If they are fending for themselves, they will usually eat all they catch, but occasionally they may catch more than they need. The surplus will be taken home to be eaten later. Some cats hide it, others leave it lying around. Only when they have kittens do they take the catch back to feed others. There is some debate as to whether pet cats see their owners as giant kittens who need feeding or whether they take their catches home just because they are surplus to requirements. The latter is more likely since they do not bring prey directly to us, but simply leave it lying around where they can pick it up later if they need to.

DID YOU KNOW?

- Some cats like to eat wool, plastic and other materials that are not part of a normal diet.
- This tendency is more common in some breeds than others. It may be connected to the unsatisfied desire to chew and consume food in a more natural form, since domestic cats are generally provided with a diet that is easily swallowed.

This is good

Consuming the kill is the last part of the hunting sequence and comes naturally to a hungry cat. Since we feed our pet cats all they need, their hunting produces food that is surplus to requirements. Uneaten prey will be brought back home where it will be played with if the cat has had insufficient opportunity to act out the full hunting sequence, or simply left on the floor.

This tastes good

Owners are often very unappreciative of their cat's hunting prowess and some owners actively scold or punish them for bringing their prey home. However, this simply confuses the cat and will not stop a behaviour that is so deeply embedded in the nature of their companion. If it sometimes appears that cats are bringing their catches back for their owners, this may be just a coincidence of them arriving home and greeting their owner at the same time.

What have you got?

This cat plays with the remains of a bird while his housemate, interested by the movement, looks on. Cats do not bring home their surplus catches to feed others unless they are females feeding kittens.

They are not protective over their spare food, unless they are still in need of it. Since, in the wild, they usually only catch just enough to eat, and go out hunting every day, they rarely have a surplus. Any left over when they have eaten their fill will be brought home and cached.

Feeding

Cats learn from an early age that humans are a good source of food. They like to eat little and often, since in the wild they would catch many small animals rather than one large one. However, owners usually prefer to feed only once or twice a day, so it is not surprising that cats are often soliciting food from them. If an owner is kind and compliant when it comes to feeding, their cat will rapidly learn the behaviour that is most likely to produce food and will display it whenever they feel the need to eat. Although owners often like these displays and enjoy feeding their cat, it is probably better for the cat if you leave dry food out for them all the time so that they can eat whenever they are hungry. Active, outdoor cats are usually good at self-regulation and rarely overeat.

Give me food

Cats that are hungry and trying to solicit food will display the tail-up greeting to you and wind themselves around your legs, rubbing their heads and bodies against you as they go. This behaviour is used in kittenhood as a greeting but cats soon learn to use it to solicit food from their owners. Since owners reinforce this behaviour with food, cats continue to use it to signal they want to be fed. Some owners mistake normal greeting behaviour for food-soliciting behaviour and this can result in too much food being given.

DID YOU KNOW?

- Cats are opportunistic feeders and, if they are sociable, it is not unusual for them to have two or three feeding stations around the neighbourhood where they can cadge a free meal from a friendly human.

I'll come up

When his owner gives the signal to come up onto the work surface, this cat wastes no time. Cats can be trained readily providing there is a big enough reward for compliance. He is not allowed up here usually but has learnt that if his owner pats the surface with her hand, he can jump up to be fed. He is young and agile and can easily make the jump.

Feed me now

He is hungry and has raised a paw to try to pull the dish towards him. His tail is raised in a friendly signal but he is focused on the food rather than his owner. Some owners do not like cats to go on kitchen work surfaces, but it is difficult to keep them off as they soon learn to scavenge for food left out.

I'm so hungry

He doesn't wait for the dish to be put down but begins eating straight away. Feeding some dry, crunchy food helps to keep your cat's teeth clean and provides the chewing exercise that cats would get if they obtained their food by hunting.

Elimination

When cats are not leaving scent messages, they like to bury their waste products so that these are not detectable. Digging a hole and covering up afterwards keeps the area clean. However, more importantly for them, it does not leave a big advertisement of their presence. Cats do this instinctively from an early age, as soon as they are able to leave the nest, if an easily rakeable substrate is available to them. They prefer to use something with a fine, clean, dry consistency in which digging is easy. This is why sand or newly dug earth is a favourite, much to the annoyance of children with sandpits and keen gardeners.

That's better

Cats like to hide away when they go to the toilet and will try to find a secluded place where they will be less vulnerable. Having dug a hole in soft soil they squat and assume a strange look of concentration. Their desire to go somewhere private results in many cats going behind a sofa or somewhere inaccessible rather than using their litter tray if it is placed in a busy thoroughfare.

I'll just cover up

Having finished, soil is raked over the hole to bury the waste products. Once this task is completed, the cat will sniff at the site as if to check their smell is reduced or no longer detectable. Having shaken their paw to dislodge any earth still clinging to it, they usually walk away from the area before settling down for a much needed grooming session.

DID YOU KNOW?

- Cats sometimes miss and deposit faeces just outside the litter tray. Placing the tray inside a larger one or using one with higher sides can prevent this.

I must go here

This cat is leaving a scent message by spraying a small amount of urine onto this bush (see page 26). The quantity of urine sprayed will be small and he will later need to dig a hole to empty his bladder. He has the same look of concentration on his face but his body posture is different as he needs to stand up to aim the urine as high as possible.

I don't like this

Cats are fastidious and do not like to eat close to their litter tray. Owners often put the food and tray close together, but this can result in a real dilemma for a cat that likes to be clean. Some cats put up with it but others take to using another part of the house instead. Cats prefer to use a clean litter tray every time so, to prevent accidents, it is advisable for owners to clean the litter tray regularly and to keep two available in case one is dirty.

Grooming

Cats spend a lot of time grooming. As well as straightening out the coat to make it a better insulator in cold weather, and helping to remove parasites, grooming can also cool a cat down in hot weather by spreading saliva onto the fur. The tongue has a rough surface with a series of backward-facing spines that act like a comb. Hairs are swallowed and coughed up later as a fur ball. Grooming also has a function in improving relationships between cats, with the less important cat making an effort to groom the more confident one. It can also be used as a displacement activity when your cat is in a difficult situation. Too much grooming can lead to bald patches if a cat is anxious or bored.

DID YOU KNOW?

- Cats have a small collarbone compared with most animals, and more supporting muscle instead. Therefore, their shoulders are not rigid and this allows them to be very flexible when the cat is grooming, twisting or pouncing on toys or prey.

Nearly finished

Cats are amazingly flexible and can reach everywhere on their bodies with their tongue except the face and head. This cat lifts a hind leg to groom the fur there. This position also allows the cat to clean the urogenital region. Cats get stiffer and less flexible as they get older and seniors often need help to clean the back of the neck and the backs of the hind legs as these areas are difficult to reach.

Now the face

The back of the head and the face are groomed by licking the edge of the front paw and using it like a cloth to pull over the ears and face. This is repeated as often as necessary to ensure a clean finish. In hot weather, a watery saliva is spread over the fur and this helps to cool the cat down considerably.

Got an itch

Back legs can scratch almost anywhere on the front of the body. Cats that have fleas sometime pull out tufts of hair on the back of the ears in irritation. This cat delicately scratches just inside his ear and on the side of the head.

This bit's tricky

Long-haired cats need help to keep their coats in good condition. Long, thick hair, especially the soft, fluffy variety found on Persians, has been specially bred for by people and cats need help if their coats are not to become matted and tangled. This ragdoll cat just about manages to keep hers under control but has trouble straightening the long bits along her back.

Claws and paws

Cats are armed with razor-sharp claws that are hooked at the ends so that they can get a good purchase on prey. They are normally kept retracted so that they stay sharp, but when the paws are used to capture prey they are protracted out of the flap of skin that covers them. Claws grow continuously and, periodically, the old outer casing needs to be removed to reveal the new, sharp claw underneath. This is done by stropping the claws down a suitable surface. Scratching is also done in strategic places so as to leave scent messages (see page 27).

Must sharpen claws

This scratching post is much too small for this cat to use properly. Cats prefer to stretch up to full height so that they can use their body weight to help drag their claws against the surface. This cat has to sit down to use this post.

DID YOU KNOW?

- Kittens are born with their claws extended. Gradually the muscles develop that retract them into a protective sheath of skin.
- The pads of the feet are very sensitive and incorporate many touch receptors.
- Some cats behave strangely just before an earthquake hits. Perhaps they can detect vibrations of the earth through their sensitive pads.

This is just right

A tree branch like this that is firmly fixed makes an ideal scratching post and will help kittens to learn to transfer their scratching to trees outside once they are old enough to go out. This is tall enough for the kitten to be able to stretch up to full height so that he can use his body weight to help bring his claws down hard enough. In fact, this kitten may be playing rather than scratching, as cats usually balance their weight over their hind legs when they scratch so that their paws are brought down vertically.

This is a good place

In the garden, cats would scratch trees or wood. However, any similar surface will do and, indoors, furniture is a favourite target. This sofa has been scratched before and smells of other cats that have been there. This young cat sniffs carefully to detect who was there and has interrupted her scratching to do so. The sofa is in a prominent place in the room and it is a good place on which to leave a scent mark.

Sleeping

Cats spend a large proportion of their lives asleep. As predators, they do not need to spend as much time eating as herbivores do and so can afford to rest for a good proportion of the day. To conserve energy and reduce the time they need to spend hunting, they prefer to sleep in warm, comfortable places. Cats seek out such places and will know all the good spots in their territory. Cats like to nap rather than spend long periods of time asleep, but if they are relaxed enough to enter into a deeper sleep they produce the same brainwave patterns as we do when we dream. During these moments, they make twitching movements of their bodies that look as though they are running and jumping, and it is easy to conclude that they are dreaming about their day's activities in the same way that we do.

I think I'll sit here

The warmth of a car engine often attracts cats and drivers are treated to paw prints over the bonnet when they return later. Kittens often crawl into car engines seeking a warm place to have a snooze and have been known to travel a great distance, having being unwittingly stowed away under the bonnet.

Ooh, lovely

People provide warmth and comfort and their laps are an ideal place to take a nap. Cats often relax completely and may even fall asleep on us, and then wake to find themselves seemingly under attack from hands moving over them. Most cats quickly realize that it is only their owner stroking them, but some cats react defensively and may scratch or bite, to the complete bemusement of their owners.

Good place for a nap

The sofa is a good place for a sleep provided noisy humans are out of the way and are not likely to sit on you. This cat is warm and comfortable and closes his eyes sleepily. He is still alert, however, and keeps his ears pricked to catch any sound that might indicate danger or that someone may be about to sit on him.

DID YOU KNOW?

- ❑ The ancestors of our domestic cats were originally desert-living animals.
- ❑ Consequently, they are comfortable lying on surfaces that are heated up to temperatures that seem too hot for us to stand.

Nice and warm here

Cats know exactly where to find a warm spot on a cool day. This cat knows that the boiler that heats the water for the house is always warm and it is her favourite place to sleep. She is so warm now that she needs to stretch out a little to cool down. She has been disturbed by the photographer and she keeps her eyes on him, but she is not concerned and soon goes back to sleep when he leaves.

Safety

Staying safe is a top priority in the wild, where there is no-one to rescue you if you are injured. In our domestic cats, the desire to be safe is still very strong. If a cat is threatened, he can choose from the four strategies of fight, flight, height and freeze to help him deal with the problem. Cats can move rapidly between these four different states, depending on how successful the one they have chosen seems to be and how great the threat is. During dangerous encounters, continuous assessment of the risks is necessary so that the appropriate action can be taken as required. This helps to keep the cat as safe as possible while maximizing the chances of dealing successfully with the threat.

What's happening?

Looking at your cat's expression can tell you a lot about how he is feeling. Apprehension and worry show in this cat's face. His ears have swivelled to find out what is going on behind him and are held slightly back, showing his concern. He is being held by someone and so has little control of what happens to him He cannot run away and this will make his fear worse.

Oh, it's scary

Kittens often use postures out of context. This kitten puts back his ears, opens his eyes wide and tucks in his tail in a mock display of dealing with an attack. He is showing conflicting signals by being relaxed enough to paw at the ball and he has raised his hindquarters as if he is getting ready to chase it.

DID YOU KNOW?

❑ The tail, ears, eyes, whiskers and body posture all help us to determine how a cat may be feeling. Studying each individually helps to identify a range of emotions that the cat may be experiencing and decipher an overall picture that is confusing and difficult to interpret.

Get away

Fighting is usually a last resort, as such close proximity to another fully armed and hostile individual is very dangerous. These two cats know each other well and are only playing, so they keep their claws sheathed as they strike out at each other. The tortoiseshell started the interaction, forcing the black-and-white cat to defend herself by striking out. She keeps her ears well back out of danger as she does so. The black-and-white cat has half-rolled so that she can use her paws more effectively and has her ears back and half-rotated showing mock aggression.

I'm not sure

This cat has just had a worrying encounter and his tail is tucked under to avoid it being injured. His fur has been standing on end to make him look bigger and is just flattening down again so is not smooth along his back. His back end is still crouched but his head is high as he looks up, trying to find a high ledge to jump onto for safety.

Body language of fear

Since cats are, by design, solitary hunters and do not readily form social groups, their signals tend to be more subtle and less demonstrative than those of pack-living animals. However, cats do have expressive body language, especially when fearful. When scared and not able to use aggression, they make their target area as small as possible by crouching down, and tucking in their ears, tail, feet and whiskers to protect them.

I don't like this

This cat is worried and crouches to make his target area smaller. He is tucking in his tail and flattening his ears to keep them safe in case an attack occurs. His whiskers are also pulled back and he stares at the aggressor confidently. Although his ears have flattened slightly, they have also rotated a little, indicating that he is a confident cat that is ready to defend himself if necessary. This gives him an angry look and the other cat that caused this look would be sensible to back down.

I'm okay

It is difficult enough to read the body language of a cat, but with Persians there is the added difficulty that the shape of their face makes it hard to read their expression. This cat looks worried, but he looks like this even when he is happy! His pricked ears and protruding paws indicate that he is quite relaxed and does not feel under threat. It is physically impossible for Persians to bring their whiskers back alongside their faces so we cannot tell how he is feeling by looking at their positioning.

I'm so relaxed

This cat is relaxed and happy and is shown here to illustrate the differences between a contented and a stressed cat. He is lying on his side with his paws extended rather than drawn in for protection and he is not ready to run. His face and ears are relaxed and his whiskers are forward. His tail is laid out and not tucked in. Hot cats have no sweat glands to help them cool down and so will often adopt postures like this one to get cooler.

I'm scared

In comparison, this scared cat is trying to make himself small by flattening himself against the table. His tail is tucked underneath out of reach and his ears are flattened against his head. His eyes look big because his eyelids are pulled back, allowing him to take in as much visual information about the threat as possible. The brightness of the light keeps his pupils narrow, but his fear response keeps them open more than usual. This cat usually meows a lot to humans and he does so now to try to get reassurance that he is safe.

Up high

For agile, athletic cats, safety often lies in getting up high away from dangers that cannot follow. Cats can jump many times their body height from a standstill and they are easily able to jump out of the way of other animals and humans that cannot do this. Having a height advantage means that they can rake with their front claws and severely injure any enemies that may try to grab them from underneath. Getting up high is often one of the first things cats will try to do if they are in danger and they will remain there until it is safe to come down.

I’m safe here

Walls and fences between boundaries represent safety from humans and dogs. Cats’ agility and wonderful sense of balance allow them to run safely along the tops of these with ease. This cat surveys his surroundings from the safety of the wall and this allows him to make safe decisions on where to go next.

Can’t get me here

This cat finds out what is happening in his world from the safety of a rooftop. Flat roofs like this are ideal lookout posts as only other cats can reach them easily. For shy or nervous cats, they provide an ideal observation point, enabling them to come down only when the coast is clear and all the dangers have disappeared.

Who are you?

Stairs provide cats with another opportunity to get up high if they are worried. This cat has found a good place from which to watch visitors. If they begin to approach, he can run up higher and make for the safety of the bedrooms. If they stay where they are, he can assess them and decide whether or not it is safe to come back down.

DID YOU KNOW?

- ❑ Providing shelves and surfaces that can be reached easily can help to give shy cats an escape route from other pets and small children that may cause them concern.

Can I get up there?

This cat is worried, as can be seen from his crouched back end, lowered tail and stiff posture. However, his head and front end are stretched up so that he can look for somewhere high that he may be able to jump to for safety.

Up trees

In the natural world, trees make ideal escape routes in case of attack by ground-dwelling predators. Cats are equipped with sharp, hooked claws that help them to run up textured vertical surfaces, and most experienced cats find it easy to climb trees. Inexperienced cats or kittens sometimes climb trees in a panic to escape from danger and then find themselves stuck since they haven't learnt how to get down. These cats are genuinely in need of rescue and may die or fall if none is available. However, experienced cats simply sit and wait for the danger to pass before climbing down. Since they may wait for a considerable time, humans often think they are stuck when they are not, only to find that they come down rapidly and easily once the emergency services arrive to rescue them.

What's happening down there?

Trees provide a natural safe haven from danger and cats are well designed for climbing and balancing on their branches. This cat surveys the dangers on the ground from his vantage point. Although he is not very high up, he is out of the way of small children and dogs in the garden below and can sit there undisturbed until they have moved on.

Going down!

Once they have learnt how to do it, coming down is as easy as going up. Some cats learn to come down bottom first with a curious hopping motion as bears do, but the headfirst approach seems more elegant and efficient, providing they are not too high up the tree to start with.

It's difficult to balance

The branch is narrow and not too safe, so this cat crouches low and uses her tail to help her balance. She is safe for the moment, but keeps her eyes open for any dangers that may be lurking.

DID YOU KNOW?

❑ Cats that live in built-up areas may not have much opportunity to practise tree climbing. If they are forced to climb to escape a danger on the ground, they may need to be rescued later.

Nothing can get me

Safe high up in the tree, this cat can get on with the pleasures of hunting and exploring. He uses his tail to help him balance and his claws dig into the branch to give him a good grip.

Hiding and keeping still

Domestic cats are small and, although they are armed with formidable weapons in the form of teeth and claws, it makes sense for them to stay out of trouble if they can. Hiding when under threat is a good strategy. The attacker may not see the cat or may not want to risk a fight when their victim is in a small, defendable space. Given that the coat colouring of our cats' ancestors would have acted as camouflage for them in the undergrowth, hiding and freezing is likely to have been a successful strategy provided they were upwind of any predators that might be passing. Domestic cats will often hide and keep still when they cannot get up high or run away in the close confinement of the home. Keeping still is often a favourite strategy for cats faced with hostility from other cats in the household. This can compromise their welfare if it prevents them easy access to necessary resources but often goes unnoticed by their owners.

DID YOU KNOW?

- Cats seem to work on the principle that if they cannot see you, you cannot see them. They will often hide their eyes from view and stay very still if they can't get completely out of sight.

You can't see me

A cat igloo bed provides a safe hiding place for this cat who is shy and worried by the people around him. Small dark spaces are attractive to cats that want to hide and, although this bed does not offer much protection, it is better than nothing. By closing his eyes and ducking his head down, he can become almost invisible and resembles a furry rug inside the bed.

You can't get me

Cats feel safe under cars but, sadly, many are run over each year because cars drive off unexpectedly. Under the car, they can shelter from dogs that are not so agile and cannot get underneath as easily, and from children who usually give up the chase once they are out of sight. This strategy also provides them with lots of escape routes should they need to change tactics and run away instead.

I'll get it

This cat may be using this log to hide in as cover for an ambush. She is crouched in a hunting pose with her ears pricked and her senses focused on something. Her forward-pointing ears and whiskers indicate that she is not worried and the pupils of her eyes are contracted because of the brightness of the day.

Aggression

Cats are equipped with claws and jaws that can inflict serious injuries. When threatened, they can use aggression to defend themselves, but this is usually a last resort when all else has failed. Before they choose this course of action, they have to weigh up the chances of success. If aggression fails to work, their opponent does not back down and they lose the fight, they risk being killed or injured. Consequently, most cats will try a whole range of other tactics to resolve the situation before resorting to violence. The strategies used for staying out of trouble vary, but if they fail, the threat appears too quickly, or they learn that these strategies don't work, they will use aggression instead.

Stay away!

This cat doesn't want to get into a fight and hisses violently and loudly to make the other cat back off. She crouches and tucks her tail in just in case her hiss is unsuccessful. Her head and face show the aggressive stance she has taken and her ears and whiskers are forward to support her hiss. However, her body stance and head carriage show that she is not completely committed to aggression and she hopes that her aggressor will respond to her hiss and back off.

Don't mess with me

This cat is really worried and is trying to make herself as big as possible to encourage her opponent to back down. The hair on her body is fluffed up and she arches her back so that there is a dramatic increase in her apparent body size. To get the most out of this posture, cats usually turn sideways on to the aggressor to show how large they are. She is hissing loudly and she stares at her attacker with her ears erect.

DID YOU KNOW?

- Most aggressive behaviour is designed to prevent conflict rather than invite it. The various signals, body postures and noises are usually intended to keep others away rather than draw them nearer.

Leave me alone

The ginger cat is not impressed by the other cat's advances and has turned his head to stare with his ears rotated so that you can see their backs from the front. The ears signal that he will be aggressive if pushed further and it is wise for the other cat to back down, as she is already beginning to appreciate. The ginger cat has a lot of confidence in his ability to deal with this situation and has not begun to show fearful defensive postures.

Don't come closer

Since the dog is not advancing, the cat can react slowly to the threat and therefore avoid triggering a sudden attack by the dog by making any fast movements. The cat has raised himself from a sleeping position and arches his back to make himself look bigger. The hair on his back is beginning to fluff up and his tail is held down out of trouble. He is ready to hiss or spit a warning in case the dog decides to come closer.

Fighting

Cats avoid fighting unless it is really necessary. However, sometimes it is the only way to settle a dispute over territory or to win a mate, and kittens will practise mock battles from an early age just in case. Since most pet cats are neutered, fights between entire males are much more rare than they would be in a wild colony. Our domestic cats have little use for fighting but will still play-fight with other familiar cats and sometimes people in the household to hone their defensive skills.

DID YOU KNOW?

- ❑ Real fights are rare and last for only a short time.
- ❑ Severe injuries are often inflicted in a real fight.
- ❑ Teeth and claws are used to good effect during fighting.
- ❑ During a fight, the defensive cat will roll over so that he can use the claws on all four feet.

I'll get you

For two equal opponents, rearing up is a good way to try to knock the other off balance to that he can be jumped on. Springing from strong back legs and using the weight of the body, each tries to knock the other one down while clawing with the front paws and biting around the ears and neck. The grey kitten has a slight size advantage but the fawn kitten has the confidence, with tail up and ears twisted to show their backs in a strong display of mock aggression.

Aarrrgh!

The fawn kitten has just reached down and touched the grey kitten on the back with a paw. Surprised and alarmed, he leaps into the air off all four feet in a self-defensive reflex action that may startle a predator enough to allow him to escape.

I'm stronger than you

These kittens are practising wrestling skills that may be useful to them later on. Adult cats would not allow themselves to be as vulnerable as this and would make sure that their heads were together so that they could protect vulnerable areas properly. Wrestling like this in adult cats is usually brief because it is potentially dangerous to be this close to another individual who is trying to bite you and rake you with sharp claws.

Take that!

Swiping with paws is a lot safer than body wrestling, especially when the body is protected as it is in this case. Cats normally begin any defensive action with the front paws and may roll onto their side to free both front paws for action. This often deters the aggressor, but if not, a wrestle may occur with both parties trying to do as much damage as possible.

Cat interactions

Although cats have a subtle body language, it is possible to look at encounters between two cats and give a good account of what they are doing and what they might be feeling at any time. Cats prefer not to get too close to each other, but sometimes it is unavoidable; for example, if two unrelated cats suddenly find themselves living in the same household. Observations of these encounters tell us a lot about the respective characters of the individuals and it may then be possible to predict whether they will eventually learn to get on with each other or whether their personalities are not particularly compatible.

Who are you?

The tabby cat finds herself too close to the ginger cat and looks up to see if there is a way out. She is a bit worried, as can be seen from her tucked-in tail, but she is confident enough to move forward to investigate. The ginger cat has not yet reacted to her presence.

Stay back

The ginger cat turns to face her with wary eye contact, ears rotating defensively, hairs up on his neck and he has shifted his weight so that he can raise his paw. This provokes a strong reaction from the tabby, who now finds herself face to face with a cat she does not know. This is too much for her and she gives a loud hiss to warn him off.

Whoops

The ginger cat seems slightly upset by this, and stares at his opponent, rotating his ears into an aggressive position. He lifts a front paw ready to strike if she comes closer. Seeing this and realizing she is very close to a potentially aggressive cat, the tabby flattens her ears, whiskers and tail to keep them out of danger and averts her gaze so as not to antagonize him further. She raises her paw just in case he decides to attack. Both cats are drawing away from each other to put more distance between them.

I'll look away

The tabby cat tries to avoid attack by deliberately turning her head as a clear signal of the slow retreat that will follow. However, her right ear and eye remain focused on him in case he attacks and she needs to change her strategy quickly.

The next generation

The urge to reproduce and pass genes on to the next generation is strong in unneutered cats. If we do not interfere, a healthy female cat with access to males and a good food supply can produce two to three litters a year. For this reason, and because tom cats are smelly and get into fights, the majority of pet cats are neutered and owners do not get to experience the full range of behaviour designed to produce and rear the next generation. Neutering not only stops more kittens being added to the numbers for whom there is no good home available, but also prevents the reproductive behaviours that are hard to live with.

Just checking

It is important for pregnant cats like this one to know their territory well so that they can find a safe place to give birth, to raise their kittens and find enough to eat during the time when there will be heavy demands on their body. If a mother feels the kittens are not safe in one location, she will move them, and it makes sense to check out likely places ahead of time.

DID YOU KNOW?

- ❑ A wild female cat's range will be determined by the availability of food. She will need enough to feed herself and her kittens.
- ❑ A wild male cat's range will be considerably larger (up to three and a half times larger) during the breeding season and will be determined by the number of potential mates within it.

Take all you need

A litter of healthy kittens needs a lot of energy. This cat is content, relaxed and well fed. Wild cats do not leave their kittens for the first few days and need to have built up fat reserves to see them through, as well as having a place in a territory where the kittens can readily find food by hunting once they leave the nest.

I'm so beautiful

Pedigree cats like this Chinchilla are 'created' by humans, who have taken control of which cat mates with which for generations. The result is a cat that is lovely to look at. However, unless the breeders take care to select for temperament and good health as well as looks, pedigree cats can harbour a genetic makeup that makes them prone to behaviour problems as well as some health defects.

Tom cats

A fully mature entire male cat will spend a good deal of his time trying to pass on his genes. To do this well, he needs to defend a sizeable territory, compete with rivals, and court females. Since all of these require large amounts of energy, tom cats often look thin and ragged with numerous scars and abscesses from their frequent fights. Smell is a very important sense for bringing together cats that normally lead quite solitary lives for mating, especially if their territories are large. The urine of a tom cat is particularly pungent so that it can act as a long-lasting reminder of his occupancy for other cats that may wander across his boundaries.

DID YOU KNOW?

❑ The Jacobson's organ in the roof of a cat's mouth allows them to 'taste' smells. To use this, they open their mouths a little and look as though they are grimacing. This is called the Flehmen response and is used particularly to find out about the reproductive status of other cats.

There's my territory

In the wild, tom cats like this one would range and roam through a large territory. Pedigree stud cats are kept confined to prevent them contracting infections from other cats and getting into fights. Although they have never experienced freedom, many of their natural desires will be frustrated and it is kinder to castrate them once their breeding days are over.

She was here!

Tom cats spend a lot of time investigating the smells left by other cats to discover the location of rivals and the availability of females on their territory. Since tom cats need to mark their territory to prevent other males from taking up residence, they begin to produce very pungent urine once they reach sexual maturity. This is frequently sprayed onto prominent places and its characteristic odour is one of the main reasons why male pet cats are so often neutered!

I'll get you

The defending cat has rolled onto his back so that all four feet can be used in defence, and he strikes out at the face of his attacker. Fighting over territory and the right to mate with the females it contains is part of the job of a tom cat, and they frequently get scratched and bitten in the process. Although there is lots of noise and posturing leading up to a fight, during which all but the most determined will back down, fights between two toms are fast and furious with each trying to inflict as much damage as possible on his rival.

Courtship

Courtship is a noisy affair with several males all attempting to mate with a responsive female once she comes into season. A female will attract many suitors, but usually one male is able to keep the others away most of the time. Once she is ready, matings are repeated frequently ten to 20 times a day for up to six days! Only the fitter, stronger males can sustain this, resulting in kittens in one litter sometimes having several fathers. This promiscuous behaviour ensures that not only are the chances of conception maximized, but also that the kittens have a variety of characters to increase the chances that some will be successful.

I'm friendly

This cat is being friendly to a human and has raised her tail in anticipation of someone stroking her. Females that are in season will adopt a posture similar to this but with the hind legs crouched so that they can be mated. This will be accompanied by rolling, purring and stretching. During the first few days of the season, the female will rub against objects more often and urinate more frequently in a variety of places to attract a mate.

DID YOU KNOW?

- ❑ Daylength and availability of food will determine when a female cat will come into season.
- ❑ Cats that live in the wild will usually have kittens in the spring when food is plentiful.
- ❑ Domestic cats with an unlimited food supply can have two to three litters a year unless they are neutered.

Get off me

After mating, as the male withdraws, barbs on the end of his penis cause pain inside the female and she turns on him in self-defence. Although this seems like a design fault, the barbs are necessary to stimulate the female cat to ovulate. This female has turned to hiss and spit at the tom with rotated, angry ears. He has turned away from her and closed his eyes to protect them in case she decides to lash out with her claws.

I'm going

The male moves off as quickly as possible in case she strikes out at him. The conflicting messages from his tail-up greeting signal and his aggressive ear position show that he is worried about her reaction and is keeping his options open as to how he might respond. The female is already beginning to recover and will soon be grooming herself ready for next time.

Kittens

Once mating has taken place, the male has no further role to play except to defend the territory where his females raise their young from other marauding males that may kill the kittens. Females give birth and raise their kittens following instinctive behaviour patterns that allow them to do so unaided, although they do get better at it with practice. In a natural colony, however, other related females will help out, acting as midwives and surrogate nurses while the mother takes a break. This cooperation ensures greater protection and survival of the young.

It's you

This cat has not seen her kitten for some time but recognizes him quickly from his smell. His presence triggers the mother cat to begin washing him with her tongue. When he was a kitten this cleansing action would keep him free from dirt and bacteria that may cause disease.

Let me clean you

While kittens are still very young in the nest, the mother will lick them to stimulate them to go to the toilet. Until the kittens are old enough to move away from the nest to be clean, she will continue to do this after each feed, swallowing their waste products to keep the nest hygienic. These kittens, however, are old enough to go to the toilet by themselves and, providing a rakeable substrate is nearby, they will already be digging holes and burying what they produce.

DID YOU KNOW?

- Kittens go limp when picked up by the scruff of the neck. This helps the mother cat when she needs to carry them to a new nest site.
- Mother cats will move their kittens if they do not feel safe or the nest site becomes dirty.

Is there enough for me?

Kittens will be dependent on their mother's milk for the first five to six weeks, until she begins to bring back prey for them to eat or until their human carers wean them onto solid food. In the wild, whole prey would be brought back to the nest, either dead or injured, to give the kittens the chance to practise their skills before they go out hunting for themselves.

Hold still

Keeping kittens clean is a vital role for mothers whose kittens may otherwise die of disease. Washing the face after eating is particularly important until the kittens have learnt the complicated process of washing this area with their paws. This kitten is happy to accept the treatment and lays back his ears so that they are out of the way.

People and cats

The relationship between humans and cats has lasted for many hundreds of years and has gone through various changes. Cats have been worshipped, vilified, used as hunters and preened on show benches. Nowadays, most cats are kept as dependent companions even though they retain characteristics that enable them to be self-sufficient. Perhaps it is this wild, independent streak in an animal capable of such affection that most attracts us to our feline friends.

I'm happy here

Cats that have been well socialized with humans during kittenhood and are treated well are happy to fulfil the role of companion and even of child substitute for childless couples. Since most cats enjoy their relationship with humans and are happy to reciprocate the greetings and affection, they fit well into our lives and make good pets. Perhaps their most successful attributes are their relative independence and their ability to exercise and go to the toilet outside, making them less of a tie than dogs that need companionship and someone to let them out on a regular basis.

Is that for me?

This kitten displays the tail-up greeting as she goes towards the child with the titbit. Repeated pleasant encounters will quickly build a strong relationship. As she grows, the kitten will learn which people can be trusted and who to keep away from. People who provide food, games and a safe place to rest will be favoured.

DID YOU KNOW?

❏ Cats are not seen as 'property' in the eyes of the law and consequently owners cannot be held responsible for damage that their cat does when on other people's property.

That's enough now

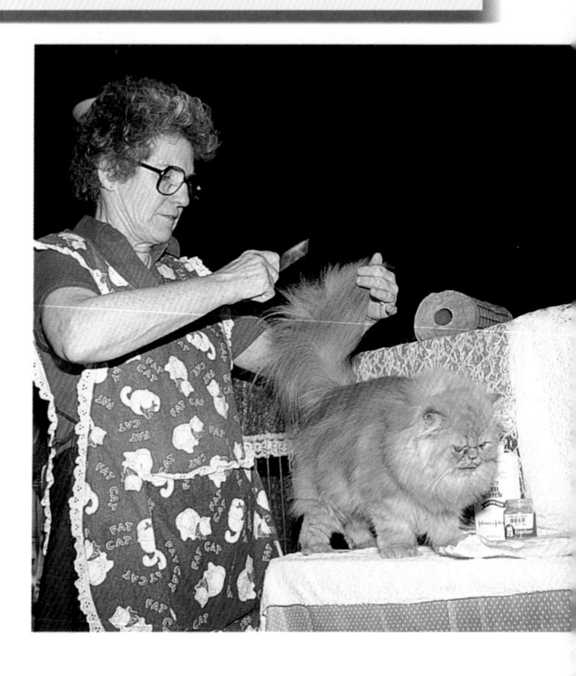

Fluffy cats like this one need a lot of grooming. However, this cat is about to be shown and a lot of extra preparation is needed. In the world of pedigree show cats, the cat's behavioural needs often come second to the pursuit of winning places. On the day of the show, cats are taken to new places, handled by complete strangers, groomed and sprayed with conditioner that changes their natural scent, and kept in small cages. Cats tolerate this because they are introduced to this life from an early age.

This is too close

This cat does not appreciate being held so close by someone he does not know. Although he has not become aggressive, he is pulling away from the child's face and looks concerned. Attempts to make a cat like you by picking him up and holding him when he wouldn't stay with you if he were free are likely to end in failure, with the cat keeping his distance in future.

Eye contact

In comparison to humans, cats are relatively small. Although they are well armed with teeth and claws, it makes sense for them to be wary of creatures our size unless they know we are no threat. For cats that have been inadequately socialized or who have been mistreated by people, staying out of harm's way is particularly important. One of the first indications that the other animal is ready for interaction is eye contact. If you are generally mistrustful of humans, it is then time to retreat. Cats stare to threaten each other and prolonged eye contact can cause all cats, trusting or not, to worry and move away.

You look friendly

This cat's curiosity causes him to investigate the new person in his territory. Since she looks away rather than at the cat, there is no need for the cat to be concerned about her. People who don't like cats often find that they are the ones that cats choose to sit on when they go to visit a cat-owning household. Their lack of eye contact allows the cat to get close, whereas people who like cats unwittingly keep them away with their friendly stare.

DID YOU KNOW

- ❑ Cats blink and narrow their eyes when they accidentally make eye contact.
- ❑ To make friends with an unfamiliar cat, blink and look away when you catch his eye.

Don't stare

Even cats that know their owners well don't appreciate being stared at. Although they get used to our eye contact when they live with us, a full-on stare can make them feel uneasy and they will usually turn their faces away to relieve the tension. Another way to make themselves feel better is to close their eyes. If they feel uncomfortable but not threatened enough to run away, closing their eyes lets them think that 'if I can't see you, you can't see me'!

Are you safe?

This toddler and kitten have just come across each other. The kitten is surprised by the child bending down and giving full eye contact and has paused with his paw in the air to decide if it is safe to proceed. This kitten has been well socialized with children and is not intimidated by the child's reaction. However, a full stare with very large eyes is enough for him to check the child's intentions before continuing with his greeting.

Safe stroking

Most pet cats will be used to being stroked and will enjoy it. However, they are equipped with sharp teeth and claws and it is sensible to try not to antagonize them during interactions. There are some areas on their body where stroking can cause them to worry. If they were not well handled as kittens, or have been badly treated or teased, it is wise to stroke the safe areas along the back, sides and tail, avoiding the head where all the sense organs are, the sensitive underbelly and delicate legs.

Do I like this?

This cat is very happy to be stroked and she kneads the carpet with her paws and purrs. She has lain down so that she can be more comfortable and she seems quite relaxed. However, this is a dangerous area to touch as the sensitive underbelly is very vulnerable. She is a very tolerant cat and if this were her owner whom she trusts, she would probably be happy for them to continue to touch her.

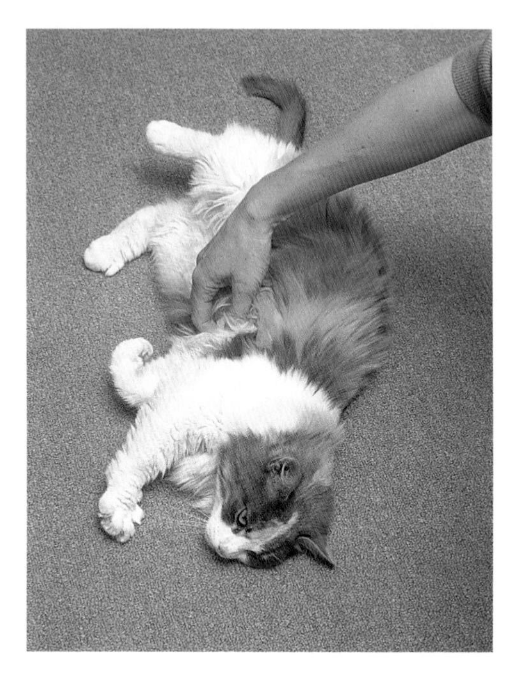

Get off!

The touch on the belly triggers a biting and scratching episode designed to get rid of the hand. This cat is very used to people and her bites and scratches are inhibited. Some cats seem to panic when they realize that they are being touched in this sensitive area and bite and scratch hard to get rid of the 'attacker'. Unless you know the cat very well, it is best to stay away from these sensitive areas.

DID YOU KNOW?

- The more gentle handling a kitten has had during the early weeks, the more friendly he will be with people.
- A kitten that has had at least four handlers during the early weeks will be relaxed and friendly with most people.

She's nice

From the neck along the back is a 'safe' area in which to stroke unfamiliar cats. In this region, no senses or sensitive areas are being interfered with and most pet cats will tolerate this. If they like you, they may then invite you to stroke their head area by trying to rub their faces and bodies against your hand. Stroking down the body and along the length of the tail is usually accepted well.

I like this

Skin glands at the base of the tail are being rubbed against the hand of the person stroking. Cats enjoy this and seem to appreciate any action that helps spread the scent onto their companion. Since some cats may have had their tail pulled by mean children or adults, it is best to stroke only the base of the tail of unfamiliar cats until you get to know them.

Training

Cats are capable of learning quickly, and they will rapidly learn that certain behaviours result in rewards such as us feeding them or opening doors for them. These behaviours can then become signals that they display regularly when they want something. It is possible for us to train them, but they will only learn if we use a variety of high-value rewards and plenty of them! Getting cross or saying 'no' never works and cats rapidly become resentful and fearful if such tactics are attempted.

Give me that food

This cat has developed a special 'signal' to let his owners know when he needs feeding. He stretches up and paws at the worktop in order to attract his owner's attention. He probably tried this when his owner was a bit slow at feeding him one day and, as it worked, he has used it ever since.

DID YOU KNOW?

- Any action that is rewarded or is self-rewarding will happen more often.
- Any action that is ignored and is not intrinsically rewarding will happen less often.

I'm coming

Teaching a cat to come when called is easy and useful. This cat has learnt that the sound of his owner calling him means food and he will go to her when he hears the sound and is hungry. His tail is raised in greeting as he runs towards her. Teaching a cat to return home for feeding can help to get him in at night when he is most at risk from being hit by a car.

Going ...

The kitten is a little hungry and is attracted to the tasty titbit being offered. As she concentrates, the titbit is lifted higher so that the kitten's weight is moved over the back legs.

... down ...

The titbit is held still in this position and with the kitten's weight over her hind legs, it becomes easier to sit down. The tail is held out for balance and the paw is raised a little to help bring down the food so it can be eaten.

... and sit

And, finally, she sits down. As soon as she does so, the titbit is given. After several sessions like this, she will begin to realize that she has to sit in order to get the titbit and a signal can be paired with the action so that, eventually, the kitten will learn to 'sit' on command.

Toys and games

Playing with our cats is rewarding for both them and us. There is a variety of toys on the market, all of which are designed to mimic the behaviour of prey. This excites the cat's hunting instincts and causes him to become active and playful. Cats play most during kittenhood and if they learn to play with toys during this time, it is likely that they will continue to do so during adulthood. Moving the toy in a way that mimics prey behaviour will result in more fun for your cat and more interest for you.

I'll get you

This kitten is attracted by the moving ball and raises her paw to swat it. Her senses are focused on the toy and she raises her tail for balance. Small, light balls like this one make good toys as they can be flicked at speed across a cat's path. Small objects that move fast from side to side make very tempting targets.

I can catch it

This toy mimics a bird as it flutters temptingly in front of this kitten. She raises a paw to bring it to the ground but her owner keeps it just out of reach. Not being allowed to catch the toy every time can be very frustrating. It is better if, in addition to sometimes missing, she is sometimes allowed to catch it and bite it as she would real prey. The game will then be very fulfilling for her.

I'll get it

If cats are enthusiastic about their toys and the games are started early enough, they can be taught to retrieve. Although this is not easy and encouragement needs to be given at exactly the right time, kittens soon get the idea. This makes the game more satisfying for the owner and the cat gets more chases. Play like this helps to further build the bond between cats and their owners.

I'm too lazy

This cat is too old and sleepy to play very much. He has played for a while and now cannot be bothered to get up and chase the toy, no matter how much he is tempted. As cats age, their desire to play and hunt diminishes slowly until, eventually, they are content to sit around for most of the day dozing.

DID YOU KNOW?

- Hiding toys in such a way that they flick in and out of cover will entice most cats to play.
- Darting movements from side to side in front of the cat are more likely to result in play than movements up and down.
- Toys that move erratically or that are very fast, then stationary are more likely to be 'hunted'.

Index

Author's Acknowledgements

I would like to thank all the people who have taught me so much about cat behaviour, in particular Sarah Heath who so willingly gave her excellent knowledge and who improved this book immensely with so many useful comments on the first draft. Thanks are also due to other behaviourists and scientists such as John Bradshaw, Sandra McKune, Peter Neville and Dennis Turner, who also shared their knowledge and made this book possible.

I would also like to thank Laura Borromeo, a wonderful rehabilitator of rescued cats in Italy, for her knowledge, her detailed comments on the proofs and her friendship. I am grateful, also, to Ryan Neile, a behaviourist with great promise for the future, for reading through the proofs and for his enthusiasm for the project.

Thanks are also due to Georgina Parker and Ryan Neile for so carefully handling and choosing cats for the photographs, and Steve Gorton for taking such beautiful pictures, often when only one chance at a shot was possible. I am grateful to the owners who loaned their lovely cats and to the Blue Cross, who allowed us to photograph some of the many cats in their care that are awaiting new homes.

And finally, I would like to thank the people at Hamlyn who made the book possible and especially Sharon Ashman, Rozelle Bentheim and Julian Brown for their help and encouragement.

Gwen Bailey

Executive editor Trevor Davies
Editor Sharon Ashman
Executive art editor Peter Burt
Designer Louise Griffiths
Picture researcher Zoë Holtermann
Production controller Edward Carter
Special photography Steve Gorton

Photographic Acknowledgements

Ardea 31 top;/Francoise Gohier 85 top.
Blue Cross/David Key 1, 23 top, 23 bottom, 36, 69 bottom, 71 top.
Getty Stone/Paul Souders 87 bottom.
Getty Telegraph/Arthur Tilley 87 top;/Maria Spann 84 top.
Octopus Publishing Group Limited/Jane Burton 57 top, 82 bottom, 84 bottom, 93 top.
Warren Photographic/Jane Burton 2, 76, 77 top, 79 bottom, 81 top, 81 bottom, 83 top, 83 bottom, 85 bottom, 92 bottom.
All other photographs **Octopus Publishing Group Limited**/Steve Gorton.